ESSENCE *of* PLACE

WATERCOLOR SKETCHES BY

MARTHA ODUM

ECOLOGICAL COMMENTARIES BY

EUGENE P. ODUM

Georgia Museum of Art
University of Georgia

FRONT COVER, TOP:
Martha Odum
Shrimp Boats, Fort Myers, FL, n.d.
Pen and ink and watercolor

FRONT COVER, BOTTOM:
Martha Odum
Monument Valley, Arizona, Navajo Tribal Headquarters, 1972
Pen and ink and watercolor

BACK COVER:
Martha Odum
Gate to Chinatown, San Francisco, 1969
Pen and ink and watercolor

ESSENCE OF PLACE

DESIGN: David Vinson
DEPARTMENT OF PUBLICATIONS: Bonnie Ramsey, Jennifer Freeman, and Hillary Brown
EDITORIAL INTERN: Kristin Collazuol
Printed in an edition of 1,000.
Printed in Spain

Support for the catalogue was generously provided by Dr. Eugene Odum. Exhibitions and programs at the Georgia Museum of Art are supported in part by the Georgia Council for the Arts through the appropriations of the Georgia General Assembly. The Council is a partner agency of the National Endowment for the Arts. Individuals, foundations, and corporations provide additional museum support through their gifts to the University of Georgia Foundation.

LIBRARY OF CONGRESS CATALOGING-IN-PUBLICATION DATA
Odum, Eugene Pleasants, 1913-
Essence of place / Eugene Odum.
p. cm.
ISBN 0-915977-39-7
1. Ecology in art. 2. Ecology. I. Title.

N8217.E28 035 2000 00-029417
910.4—dc21 CIP

table of CONTENTS

ACKNOWLEDGMENTS	*William U. Eiland*	PAGE *two*
INTRODUCTION	*Eugene P. Odum*	PAGE *four*
ESSENCE OF PLACE	*Eugene P. Odum*	PAGE *six*

acknowledgments

Eugene and Martha Odum had an extraordinary love affair that lasted for over fifty years. Their marriage was one based on the concert of mutual respect and its attendant harmony of shared interests. Martha, as an artist, had a special talent to record the essence of a place with a minimum of detail; Eugene, as a landscape ecologist, looks at her landscapes and comments on the ecological principles that they remind him of. While Eugene Odum's concerns have perforce been global, Martha's were more often than not local, yet not restricted to the community of Athens or the state of Georgia. Her quest for the perfect expression, through her art, for the defining characteristic of place made her, albeit briefly, at home in Big Sur, in Venice, or in the Far East. She attempted to remove from the grand sweep of the natural the detail that signifies the whole. She would have perhaps approved of Sir Thomas Brown's observation in the *Religio Medici*: "All things are artificial, for nature is the art of God." Eugene, on the other hand, has had to extract from the considered, vast accumulation of scientific fact a bigger picture, one that must take into account man's place in that landscape. He, perhaps, would agree with John Dryden's verse from the *Annus Mirabilis*: "By viewing Nature, Nature's handmaid, art,/Makes mighty things from small beginnings/grow." The present volume is a record not only of this marriage of science and art but also a testament to their lifelong romance.

When Eugene Odum suggested this volume to us, we embraced his idea enthusiastically, and he gave us the means to accomplish its publication. We are grateful to him and to Martha, to whose memory we dedicate this book, for their devotion to the museum and its mission of celebrating the fine arts. I am thankful that Bonnie Ramsey and Jennifer Freeman of the museum's staff understood implicitly Dr. Odum's plans, and they, in turn, appreciate the chance to share his vision and, once again, to help preserve Martha's legacy. Hillary Brown, one of our interns at the museum, was of great help, and I thank her. All of us at the museum acknowledge with great appreciation the assistance of Professor Edward Lambert of the Lamar Dodd School of Art. It is so much more than a pleasure for all of us to work with Eugene Odum: it is an honor of which we can only hope that we are worthy.

William Underwood Eiland
Director
Georgia Museum *of* Art
University *of* Georgia

...Martha, as an artist, had a special talent to record
the essence of a place with a minimum of detail;

Eugene, as a landscape ecologist, looks at her landscapes and comments
on the ecological principles that they remind him of...

introduction

Art and ecology have been a partnership in the Odum family for more than fifty years. Martha, in addition to being a gifted designer and craftsperson, was a landscape painter who was able to capture the essence of place in her watercolors. Along with our son and my brother, I am a landscape ecologist who studies the "essence of function" of landscapes as living, pulsing ecosystems.

Everywhere we traveled, Martha carried a large pad of watercolor paper in her suitcase for on-the-spot painting. She also carried a 5 1/2 x 8 1/2-inch rag paper sketch pad, a small metal box of watercolors, and a vial of water (also a vial of rubbing alcohol to use when the temperature was below freezing) in her bag. Even when she had only half an hour at some scenic spot, she was able to complete a painting that truly captured the essence of that place. Sometimes she used these smaller sketches as a basis for larger pictures that she painted in her studio using acrylics as well as watercolors.

It was her wish, as specified in her will, that her larger pictures, other than those acquired by museums, be distributed to relatives and friends who would hang them in homes and offices where they could be seen and enjoyed. Several of these large paintings are included in *Martha Odum: Watercolors*, published by the Georgia Museum of Art in 1997 and available in the Museum Shop. I have retained 150 or so small sketches mounted in four photo albums. I have selected, with the help of Jennifer Freeman, Bonnie Ramsey, and Hillary Brown of the Georgia Museum of Art and Edward Lambert of the Lamar Dodd School of Art, seventy-three of these for this book. I have had no difficulty with coming up with an ecological message and/or an interesting anecdote for each of the selections. So, this is a book of landscapes with environmental vignettes.

Eugene P. Odum
Callaway Professor Emeritus
Institute *of* Ecology
University *of* Georgia

*...Martha, in addition to being a gifted designer and craftsperson,
was a landscape painter who was able to capture the
essence of place in her watercolors.*

*Along with our son and my brother, I am a landscape ecologist who studies the
"essence of function" of landscapes as living, pulsing ecosystems...*

Essence *of* Place

Before the stock market crash of 1929 and the subsequent Depression, Martha's parents had a summer house on Lake Geneva, Wisconsin, about half a day's drive from their home in Wilmette, on the north shore of Chicago. When Martha was in grade school and high school, her

a road remembered

family made many trips there. In 1938, Martha painted this picture from her memory of the drive through the rural landscape, which in those days included the two-lane highway, small, neat farm fields, natural roadside vegetation, and forested hills.

Today, this landscape has greatly changed. Highways are wider and straighter, farm fields are fewer but much larger, and industrial development and suburban sprawl are rapidly spreading into the area. It is becoming more evident each year that we must replace haphazard development with some kind of reasonable land-use planning (or "smart growth," as some people call it) that will include preserving greenbelt buffers along streams and roads and between areas of intensive development. These buffers will not only make the landscape more attractive visually but will also preserve biodiversity, slow the spread of pests and diseases, and contribute to the maintenance of air and water quality as well as nature's other life-support services.

OPPOSITE: PICTURE PAINTED FROM MEMORY AFTER RIDE THROUGH HILLS OF WISCONSIN NEAR BARABOO IN 1938 (1938)

Martha and I spent our first year together (1939-40) in the little Dutch village of Rensselaerville, located in the foothills of the Catskills, thirty-five miles southeast of Albany. I was the first resident naturalist at the then newly established Edmund Niles Huyck Preserve, now a

rensselaerville, NEW YORK

well-endowed research and conference center. That first year, I did an inventory of the vegetation of the preserve and a detailed study of the annual cycle of the black-capped chickadee. At that time, most bird life-history studies focused on the nesting season; I wanted to find out what birds do during the rest of the year, when they are not nesting, and how a tiny, non-migratory bird survives the winter, including information on flocking and roosting behavior and physiological adaptations, such as winter fat deposits.

We essentially had a long honeymoon in the deep snows of winter (this was before global warming). In the winter, we lived in the right wing of the house shown in Martha's picture; in the summer, we moved to a cottage on a pond in the preserve. Martha painted more than fifty pictures that year, one of which (a view of the village street in spring) was purchased by the Albany Museum of Art for its permanent collection. Most of the others she sold to people who had summer homes in the vicinity.

Later, she sometimes regretted selling so many of her early works, but in her will, she specified that her remaining large paintings be distributed free to homes and offices where they would be seen and enjoyed over the years.

opposite: Our First Home, Rensselaerville, NY (1939-40)

If you want to see Scandinavia in all its Viking glory, start at Bergen, Norway, take a ferryboat up one of the fjords (see sketch above), and stay overnight at a hotel with a magnificent view, such as the one at Flåm in Martha's watercolor. The next day, take the

scandinavia

train over the high pass into Sweden, visit Stockholm and vicinity, including the university city of Uppsala, then head north to the land of the midnight sun, where the reindeer roam.

The Swedes are among the most environmentally aware people on the planet, and their scientists are leaders in ecological thinking and research. Recently, Swedish ecologists came up with a new concept: the Ecological Footprint, a definition of the amount of environment outside a city required to provide food, water, waste treatment, air purification, forest products, and other natural resources. The footprint of a modern industrial city is larger than most of us would estimate. In a recent paper published in the journal *AMBIO*, a team of Swedish ecologists calculated that the area of rural and natural land and water required to provide these life-supporting goods and services for the twenty-nine cities that ring the Baltic Sea is between five hundred and one thousand times the area occupied by the cities themselves.

All the landscapes illustrated in this book are not only interesting and beautiful scenery, but also ecosystems that provide the largely non-market goods and services that we need to survive. As we enter the twenty-first century, we must try harder than ever to preserve the quality of these environments.

OPPOSITE: VIEW FROM HOTEL TRIETEM, NORWAY (1958)

In 1958, we visited the Isle of Cumbrae, a small island of only a few square miles off the western shore of Scotland in the firth of Clyde. Yes, there is a marine station on the island, so we had to visit it, especially since I had become acquainted with the director on his recent visit to the United

isle of cumbrae, SCOTLAND

States. The red sandstone rocks covered with green seaweed and the pink sand beaches backed by green hills made for an unusual and beautiful landscape. The marine station is well-sited in the island's main town of Millport, with access to a variety of marine environments (rocky, sandy, and muddy shores, a large estuary, and deep water with soft mud, gravel, and rocky bottoms).

As might be expected in Scotland, there was a golf course in the hills above Millport. On the last day of our visit, my son and I decided we would play nine holes of golf in the morning, before the noon ferry back to the mainland. As the day grew warmer, I took off my sweater, laid it on the edge of one of the greens, and forgot about it. Later, when we were boarding the ferry, a man came running up with the sweater. Hospitality and goodwill like that make one optimistic about humankind worldwide.

Millport, Isle of Cumbrae,
Scotland, Red Sandstone and Beach
(1958)

Although England does not have as rugged a landscape as Scotland, it does have its Lake District (a national park) and a good-sized range of mountains in northwestern England. Lake Windermere is one of the larger bodies of water in the Lake District, and a biological station has been operating on its shore for many years. While we were visiting the station, which has guest rooms for visiting scientists, Martha painted this watercolor.

lake windermere, CUMBRIA

From afar, the lake looks pristine, but as is the case for most lakes, it is mildly polluted and enriched with run-off nutrients; this has resulted in some changes in its organisms, such as an increase in green plants that float on top of the lake (phytoplankton) and a decrease in those on the bottom.

What we now call holistic or ecosystem ecology (the study of how a whole landscape unit, such as a lake or forest, functions) began early in the twentieth century with the study of northern European and North American lakes. The "pulse of life" or "community metabolism" involving carbon dioxide uptake, oxygen production, and nutrient cycling is much easier to measure in still water than on land or in running water. Out of such measurements comes one of ecology's most important concepts, the P/R ratio: the ratio between the amount of energy produced by the ecosystem and the amount of energy required to maintain the system, i.e., respired by it. When P is greater than R, the system grows; when P equals R, growth stops unless there is an energy subsidy from outside. This latter scenario is the case with cities, which consume much more energy than they produce and are thus dependent on imports, especially fossil fuels (nature's energy stored in the past). No system of man or nature can persist if the energy necessary to maintain it exceeds the energy available; overall, R cannot exceed P.

opposite: Lake Windermere, Cumbria (n.d.)

In 1955, I was a delegate to the "Atoms for Peace" conference in Geneva. The enthusiasm for the projected atomic age was very high. One speaker predicted that by 1975, electricity generated by atomic energy would be so cheap that we would not bother to meter it; you would pay a small fee to use as much as you wished. Well, 1975 has come and is long gone, and the future of atomic energy's peaceful uses is still uncertain.

While in 1955 I was skeptical about the overall benefits of this new technology, I decided that I needed to learn more about it, especially in relation to the environment. In addition to continuing to work on the use of radioactive tracers to map food chains at the Atomic Energy Commission's (AEC) Savannah River site, I applied for and was granted a National Science Foundation Senior Fellowship to spend a year visiting other research sites.

the horse heaven hills of EASTERN WASHINGTON STATE

One of the sites we visited in the spring of 1958 was the AEC's Hanford Plant near the tri-cities of Richland, Kennewick, and Pasco in eastern Washington, where the Yakima and Snake Rivers flow into the Columbia River from the east and west, respectively. It was here that the biological magnification of radioactive isotopes was first documented. Scientists found, for example, that even a very small amount of radioactive phosphorus leaked into the river ends up concentrated in dangerous levels in wild Canada goose eggs.

Martha enjoyed the wide open spaces of eastern Washington as a contrast to the "green jungle" we live in in Athens, Georgia. The sketch reproduced here is a view across the Yakima River into what are known locally as the Horse Heaven Hills, an area covered with desert and grassland vegetation where ranch horses and wild mustangs graze.

OPPOSITE: YAKIMA RIVER, HORSE HEAVEN HILLS IN THE DISTANCE, EASTERN WASHINGTON (1958)

For reasons I do not remember, our trip to the Olympic Peninsula in April 1958 was very hurried, with little time spent in any one place. This sketch of a small rocky outcrop island off Ruby Beach shows what Martha could do in about five minutes: a sort of abstract essence of place.

olympic peninsula, WASHINGTON

The Olympic Peninsula is an area of exceptional variety and beauty, with its wild sea coasts and the temperate rain forest of Douglas fir and hemlock in the interior. Much of this beauty is preserved in the Olympic National Park and adjacent Olympic National Forest, both of which are near Seattle.

opposite: Ruby Beach, Olympic Peninsula, WA (1958)

This sketch of a small rocky outcrop island off Ruby Beach shows what Martha could do in about five minutes: a sort of abstract essence of place.

Both the artist and the ecologist view villages and small towns in third world countries as much more attractive and interesting than the large cities there, where the population of poor people continues to grow too fast. Thus, the landscape of a town like Taxco, with church spires that are the tallest constructions in town, the town square where people are in no great hurry, and the open air food and craft markets, is much more paintable than the crowded and polluted landscape in and around Mexico City. One has the feeling that humans living in villages are much more in tune with their environment than in the third world megacities.

taxco, MEXICO

The question, then, is: why are so many people leaving the tranquil countryside for the city? Partially to seek a better economic future, and partially because industrial technology is forcing them to move. Paul Gray, a former president of the Massachusetts Institute of Technology, recently wrote, "The paradox of our times is the mixed blessing of all technology." The bright side of industrial agriculture is the increased yield obtained per unit of land, but the dark side is that small farmers cannot afford the machinery and chemicals required for large-scale industrial farming. They are thus put out of business and forced to emigrate to cities, where they instantly become poor and have to buy rather than grow their food. A possible way to reduce this human flood is to promote new cash crops that are labor-intensive but have higher unit values, such as vegetables and flowers. This method allows the small farmer to survive in Holland (see page 76).

OPPOSITE ABOVE: Taxco, Mexico #1 (1960)
OPPOSITE BELOW: Taxco, Mexico #2 (1960)

M.ODUM

M.ODUM

In 1962, I was invited to spend two months (April and May) on a lecture-teaching tour in Japan, funded by a grant from the Japanese Ecological Society and the Japanese Society for the Advancement of Science. Martha, our son Bill (then twenty years old), and Martha's mother, Bernice, came along. We followed the cherry blossoms from the southernmost island of Kyushu to Sendai in the north, visiting all eight of the Imperial universities and many marine labs. Martha

nagasaki, JAPAN

painted up a storm, including a number of large finished pictures and numerous essence-of-place sketches, several of which are included in this book.

At Nagasaki, where the second atomic bomb was detonated, we thought that the people might resent Americans visiting the city and its museum of the Holocaust, but this proved not to be the case. The Japanese people are very resilient and pragmatic, taking the view that if they could not beat us in battle, they would beat us in automobile manufacture and other business.

Fortunately, the environment is also resilient, and both natural processes and human efforts continue to restore life and Japanese beauty to this seaside city.

opposite: Nagasaki, Japan (1962)

Japan is a nation of islands, four big ones and hundreds of small ones. The western coasts along the Sea of Japan, between Japan and Korea, are especially scenic, with numerous small islands and rocky outcrops that were thrust up from the bottom of the sea during past volcanic activity. As so beautifully shown in Martha's painting, these shallow southern seas are tropical in coloration, with patches of deep purples, greens, and blues due to the variety of underwater seaweeds and reefs. These

shirayama, SEA OF JAPAN

colorful seas sharply contrast with the clear, cold, light blue or gray northern seas.

While Martha was working on this picture, I visited one of the country's numerous marine laboratories, many of which have an aquarium out front where people pay to watch fish and other sea creatures. Profits from these aquaria directly support the marine research that is vital for the country. The Japanese must fish the world seas and promote aquaculture at home to obtain enough protein to sustain their dense population.

As the old saying goes, "you are what you eat." Thus, American culture revolves around the steak, the cowboy, and home on the range, while Japanese culture is much influenced by fish, fishing, and fish watching. Once a year, the Japanese celebrate Children's Day with fish-shaped balloons hoisted on tall poles, one for each child and grandchild in the extended family.

As the old saying goes, "you are what you eat." Thus, American culture revolves around the steak, the cowboy, and home on the range, while Japanese culture is much influenced by fish, fishing, and fish watching.

Shirayama, Japan (1962)

The Japanese hot tub is, philosophically, entirely different from the Western bath: you clean yourself before you bathe. When you get out, the water should be just as clean as when you went in, clean enough to be recycled for another bath. In Japanese inns, there is a drain in the bathroom floor: you sit on a wooden stool and clean yourself with warm, soapy water, or in more up-to-date inns, you take a shower before getting into the tub.

view from hotel kanden club in the JAPANESE ALPS

Several hot tubs in the Hotel Kanden Club in the Japanese Alps (where the 1998 Winter Olympics were held, in Nagano) have a view of the mountains through plate glass windows. Martha painted the picture on the opposite page while in the tub, up to her neck in very warm water, an ideal, relaxing environment in which to do watercolor sketches.

OPPOSITE: SHINANO OMACHI, JAPAN,
VIEW FROM HOTEL KANDEN CLUB (1962)

Almost every family of warm-blooded vertebrates (birds and mammals) has at least one species that is aberrant, or completely different in appearance, habitat, or lifestyle from its close relatives. For example, most woodpeckers are black and white with some red around the head, and most feed on insects in bark and dead wood in trees. The flicker, however, is a brown woodpecker with a spotted breast like a thrush, and it spends most of its time on the ground using its sticky tongue to feed on ants. The evolutionary explanation is that by occupying a different "niche," the aberrant species avoids competition with its close relatives.

the japanese MONKEY

Most monkeys are slender, with long tails and arms, and swing through the trees of tropical forests. In contrast, Japanese monkeys live in northern Japan, where winter brings lots of ice and snow. They are heavyset and spend most of their time at ground level in rocky mountain country. Most that survive today live in parks or around shrines, where they are protected. In such places, they become tame and spend much of their time trying to steal picnic lunches. Often, tourists feed them like pigeons in a park.

The Japanese monkey in Martha's sketch was sitting on a large rock and seemed to be meditating like the man in Auguste Rodin's famous statue *The Thinker*.

OPPOSITE: JAPANESE MONKEY, BEPPU (1962)

mount fuji, JAPAN

Mt. Fuji is one of perhaps a dozen or so of the world's best known mountain peaks. Mt. Everest, the world's highest, Mt. McKinley in Alaska, North America's highest, and Mt. Kilimanjaro, in East Africa, are others that come to mind. All of the world's highest peaks, even in the tropics, are covered by snow and ice throughout the year. These frozen caps are not lifeless, as might seem at first glance, but support well organized microbial communities, including green algae, and sometimes a few hardy insects living on, in, or under the ice. These biotic communities obtain their food from two sources: photosynthetic conversion of solar energy and organic debris (detritus) blown up by wind from the forests below.

Mt. Fuji is a major tourist attraction in Japan, and there are many tourist hotels in the valleys below that have a view of the mountain. The problem is that the mountain is covered with clouds much of the time, so a tourist may have to wait a day or two to get a view of the peak. When we checked into the Fujiview Hotel in the afternoon, the peak was completely covered by clouds. Martha had a premonition that the clouds might lift early in the morning, so she was out on the viewing porch at dawn, and sure enough, there was the peak in all its glory. By the time she had made three or four watercolor sketches, the clouds had returned. The sketch shown on the right was painted as the clouds began to return.

OPPOSITE: MT. FUJI, JAPAN (1962)

The Columbia Glacier near Anchorage, Alaska, is a popular tourist attraction. Martha sketched it from the deck of a tour boat early in the season (June) when there were few people on board. When the captain saw her painting while standing at the rail, he

the columbia GLACIER

brought out a chair and table for her to use as an easel. As I remember, she completed three or four watercolor sketches, one of which she gave to the boat captain, during the forty-five minutes that the boat was in view of the blue hunk of ice that protrudes into the bay; two of these are reproduced on the opposite page.

While the amount of water on earth has remained constant in recent geological ages (millions of years), the amount that is frozen has ranged from not much in past tropical ages to a great deal during past ice ages, when the polar ice caps extended far beyond their current reach. During the last half of the twentieth century, the polar ice caps and mountain glaciers (including the Columbia Glacier) have been in slow retreat due in part, at least, to human activities that promote global warming. If all the ice melts, the sea level will rise and flood the coastal lowlands and river deltas where the majority of the earth's population resides. So, preservation of the Columbia Glacier is just as important for our future as preservation of the rain forests.

OPPOSITE ABOVE: COLUMBIA GLACIER, ALASKA #1 (1962)

OPPOSITE BELOW: COLUMBIA GLACIER, ALASKA #2 (1962)

The scenery along the southern coast of Alaska rivals that of Norway's fjord country. The Valdez Narrows is one of numerous steep-walled bays and dead-end inlets along this coast. Early explorer Captain James Cook tried unsuccessfully to find a passage through the

valdez narrows, ALASKA

Narrows to the Bering Sea. It is also a region of monster tides called "tidal bores." Water piles up in the narrow and shallow dead-end bays on the incoming tide, then comes roaring out down the channel as a ten-foot or higher wall of water on ebb tide. Obviously, it is no place to be in a small boat.

Martha's sketch was one of a series painted on a tour boat, cruising slowly along the coast, the same boat from which she painted the Columbia Glacier (see previous entry). At this season, there was still some snow at low altitudes. The evergreen spruce trees that grow on the slopes are short because of the brief growing season, but they may grow in dense stands, as seen on the right.

Valdez Narrows, Alaska (1962)

Seashores and mountains especially inspired Martha to paint. Her watercolor of the panoramic view of the Rocky Mountains as seen from the road that leads to the small town of Gold Hill, Colorado, captures the essence of zonation of vegetation that is characteristic of all mountain

the arapahoe glacier, COLORADO

ranges. We see the temperate semi-arid grassland in the foreground, then the bands of dark green northern conifer forests higher up, and still higher, the treeless alpine meadows, so distant that individual plants cannot be distinguished. The range is not high enough at this point for year-round snowcap, but there is a small glacier in the sheltered pocket between two peaks. For a commentary on the ecological importance of glaciers, see page 32.

In high, steep mountains, you can go from a temperate climate to an arctic one in a few miles when traveling up the slope. In flatlands, you would have to travel hundreds of miles to experience the same change of climate.

About twenty percent of the earth's land area is mountainous and about ten percent of the earth's population lives in mountains, more in Europe and Asia than in North America. Each of the continents has one or more high mountain ranges. No matter where you live, mountains are vital to maintaining the quantity and quality of water; almost all of the world's great rivers originate in mountains, where rainfall and snowfall are usually greater than in the plains below.

About one-third of the solar energy received by the earth is dissipated in driving the water cycle that involves evaporation and desalination of water from the sea, formation of clouds over the continents, and rainfall and downhill flow that return the water to the sea. This water cycle (or recycling) is one of the vital services of nature, and it's free as long as there is plenty of water, but as shortages develop, water will become increasingly expensive.

View of Arapahoe Glacier from Road to "Sunshine" and Gold Hill, Colorado (1964)

colorado SAND DUNES

Sand dunes in the middle of Colorado, miles away from the shore of any sea or lake—how can this be? The answer is wind. Strong winds that blow constantly in one direction in dry valleys can build up sand dunes, like those in the deserts of the Middle East, that are very similar in appearance and plant life to beach dunes.

We were on our way home from a visit to the West Coast when we decided to go by the Great Sand Dunes National Monument, which was featured on the road map of Colorado. To Martha, sand dunes represented wild, untamed nature, constantly changing in form and open to wind and sky. She also painted several pictures of coastal dunes at Sapelo on the Georgia coast (see page 59).

Wind is an indirect form of solar energy since it originates from differences in temperature and air pressure resulting from the sun's heat. In recent years, interest in tapping some of this energy with "forests" of large windmills has developed in areas where strong winds blow most of the time. Wind power, combined with photovoltaic panels that convert sunlight directly into electricity, can contribute to a mix of renewable energy sources that will replace non-renewable fossil fuels as they are used up in the twenty-first century. The difficulty with solar power is that, compared to coal, oil, and natural gas, it is a very dilute, or low quality source that has to be concentrated to drive machinery. Thus, sunlight will not run your car unless it can be upgraded to a higher quality form such as electricity or hydrogen, a costly process. There is no such thing as a free energy lunch in our future.

OPPOSITE ABOVE: Sand Dunes National Monument, Colorado (1964)

OPPOSITE BELOW: Sand Dunes, Colorado (1964)

M. ODUM

Some of the most colorful landscapes in the United States are in southern Utah and the Monument Valley–Four Corners region, the only place in the country where four states (Utah, Colorado, New Mexico, and Arizona) share a common border. Martha painted in this region

the first snows of september in the LA SAL MOUNTAINS, UTAH

on several occasions between 1964 and 1993. *First Snows of September* is one of her many watercolors of this area, several of which are reproduced in this book.

In this view, bands of red sandstone in the foreground contrast with the blue mountains in the background. Winter comes early in these mountains. Usually by late September, snow has appeared on the high peaks, as when this picture was painted, in 1964.

opposite: First Snows of September, La Sal Mts., Moab, Utah (1964)

M. ODUM

The eastern Appalachian Mountains are much older, geologically speaking, than the western Rocky Mountains. Heavy rainfall over the ages has worn them down, resulting in rounded mountaintops rather than peaked ones like those in the Rockies (see page 37).

blue valley overlook, NEAR HIGHLANDS, NC

Highlands, North Carolina, the highest incorporated town east of the Mississippi at an average of 4,118 feet, is located on a plateau surrounded by dark green forested and rounded peaks. Many affluent people from Florida and Atlanta, Georgia, have summer homes there. We spent several summers at Highlands when I was engaged in research at the Highlands Biological Station. The diversity of birds, small mammals, salamanders, and fungi makes this high rainfall region attractive to ecologists. This is also where our son, Bill, when he was three years old, caught his first trout and decided he wanted to become an ichthyologist, though he could hardly pronounce the word.

Martha's objection to Highlands was not only its wetness, but also the lack of panoramic views from the town itself. One had to drive ten miles west to an overlook to get a sense of the essence of place of the whole landscape. The high humidity often creates a blue haze that hangs over the mountains and valleys in this region, which is why the vantage point is called the Blue Valley Overlook.

Blue Valley Overlook,
Near Highlands, NC (ca. 1964)

In summer, the shores and coves of Cape Cod are replete with flowers and sailboats. Martha not only loved to paint flowers and boats, but she could also raise flowers and sail boats. As a teenager, she learned how to maneuver sailboats and how to race them on Wisconsin lakes, where her parents had their summer cottage. She even won some races.

The flowers in the foreground of this painting are part of a large, privately maintained seaside garden known as "The Little Indian Garden" because of a wooden Indian that stands at the entrance (not shown in this picture).

little indian garden, SIPPEWISSET, CAPE COD

During the four summers from 1968 to 1971 that we spent in Woods Hole on Cape Cod, Massachusetts, when I taught the summer course in marine ecology at Marine Biological Laboratory (MBL), Martha was in seventh heaven in terms of painting opportunities. At one of her open house exhibitions at Woods Hole, a small boy, after carefully looking at all the pictures, asked, "Mrs. Odum, do you ever paint any pictures that don't have water in them?" The answer, of course, is yes, in locales where water is not a major feature of the landscape, as the samples of her work selected for this book indicate.

Little Indian Garden at Sippewisset, MA
(1966)

In Bermuda, white tile roofs are a distinctive feature of the landscape. To make up for low rainfall, water running off the spotlessly clean roofs is collected in basement cisterns. When you build a house in Bermuda, you first lay out a large swimming pool as a reservoir,

the white roofs of BERMUDA

then construct the house over it. Martha's picture captures the "essence of place" of this adaptation to the local climate.

Islands are especially interesting to ecologists; on one hand, the isolation reduces the number of species present, but on the other hand, the same isolation enhances the evolution of new genetic varieties and species. An ecological principle known as "island biogeography" holds that both the size of an island and its distance from the mainland affect the diversity of land flora and fauna.

Small islands like Bermuda that are a long way from the mainland have fewer species of native land plants, birds, mammals, and insects. With reduced competition, many island species occupy a wider ecological niche than their relatives on the mainland. For example, a bird that is restricted to a forest habitat on the mainland may expand into open country on an island in the absence of a competitive open-country species. Humans have a habit of introducing species to islands, sometimes to fill what they feel is an open niche and sometimes just because they want to have the species around, as with mainland North American cardinals, which have been introduced and are now thriving in Bermuda and Hawaii.

Bermuda (1966)

Here we have a subtropical summer sky with large, white, billowy clouds and a thunderstorm popping up in the distance. When our son was completing his Ph.D. work at the University of Miami Marine Laboratory, we often stayed at a hotel on Key Biscayne, south of Miami, a tranquil place with lots of palm trees and great views of sea and sky.

a south florida SUMMER SKY

Cloud cover, of course, varies with region and season and is increasingly affected by human activities. Clouds can either heat or cool the landscape, depending on what they contain other than water. Over cities, clouds often contain dust and fine particles (aerosols) that reflect the heat of the city downward, making it warmer and sometimes wetter than the surrounding countryside. On the other hand, clouds also reflect the sun's heat back into space, cooling the earth's surface. How this uneasy balance between heating and cooling will play out is difficult to predict in any one place.

Perhaps the worst thing we do to clouds is to create acid rain. When fossil fuels are burned, especially in coal-fired power plants, sulfates and nitrates are discharged into the air. These react with water in the clouds to produce sulfuric and nitric acids, which fall to earth as acid rain. Trees on hills and mountains that are bathed in acid clouds for long periods of time turn brown and die. In some northern lakes that lack alkaline buffers, the acidification is so bad that no fish can survive. We have the technology to eliminate acid rain by either gassifying or liquefying the fuel and burning only that acid-free gas or oil. Such procedures, however, might add a few cents per kilowatt to your electric bill, so power companies won't use this technology until enough people and politicians understand the situation and demand that it be done.

View North on Beach,
Key Biscayne, FL (1967)

It was in these Chiricahua Mountains, in the southeast corner of Arizona near the Mexican border, that Geronimo the Apache Indian chief and his men held out against the U.S. cavalry in the nineteenth century. Today, the mountains and canyons are a favorite attraction for tourists,

cave creek canyon, ARIZONA

especially bird-watchers, who come to see the unusually large number of species of hummingbirds and tropical species such as trogons, colorful parrot-like birds.

During the two days that we explored the Chiricahuas, in the summer of 1968, Martha showed no inclination to paint. She was content to be a tourist, enjoying the views and looking for trogons. It was not until we were leaving in our rented car and paused at the canyon entrance that she suddenly whipped out her pad and began painting. I think it was the depth and diversity of landforms as seen from the adjoining desert that challenged her.

Interestingly, the shrub desert seen in the foreground at the entrance of the canyon has a larger variety and greater numbers of small rodents such as kangaroo rats, picket mice, and ground squirrels than are to be found in a moist, eastern forest. Why? Seeds. Desert plants survive by producing many seeds that remain dormant in the soil until it rains hard enough to provoke germination. Accordingly, there is a surprising abundance of seed-eating animals and insects in dry deserts.

ENTRANCE TO CAVE CREEK CANYON,
PORTAL, ARIZONA (1968)

Martha was skilled in capturing the essence of place of natural landscapes with a minimum of brush strokes, but she was also proficient in more detailed paintings of human-made landscapes, such as this one of the entrance to Chinatown in San Francisco. She could draw as well as she could paint. Compare this sketch with the one of the Golden Gate Bridge on the next page to see contrasting styles.

chinatown, SAN FRANCISCO

San Francisco is perhaps everyone's favorite city to visit, not only because of its mild climate and spectacular natural environment, but also because of the diversity of cultures that, in general, tolerate one another. One way for two very different cultures to coexist is for one to be enclosed as a gated community within the other. In the case of Chinatown, the gate is always open for people to pass in and out, in contrast to many suburban gated communities. Whatever our adaptation to human multiculturalism, the overall relationships between humans and the environment need to be improved greatly. When it comes to pollution, urban sprawl, wasteful consumption, crime, and poverty, San Francisco equals other large cities.

OPPOSITE: GATE TO CHINATOWN, SAN FRANCISCO (1969)

opposite: Golden Gate Bridge, San Francisco (1969)

the golden gate BRIDGE

Perhaps the two best known bridges in the U.S. are the Brooklyn Bridge and the Golden Gate Bridge. Martha has painted both as well as many smaller ones in America, Europe, and Japan. The fog in the San Francisco Bay area often makes the Golden Gate Bridge seem to appear and disappear, like a giant mechanical ghost.

Bridges not only connect land areas separated by water or canyons, they are also works of art. The word "bridge" is also often used to designate the joining of two cultures or disciplines. In the late 1960s, English scientist and writer Sir Charles P. Snow wrote a book entitled *The Two Cultures* in which he expressed concern with the lack of communication between the humanities and the sciences in academe. He suggested that a "third culture" was needed to bridge the gap. In 1997, I published a commentary in the *Bulletin of the Ecological Society of America* (78: 234) suggesting that ecology could contribute to this third culture because it is an example of the large-scale, holistic, or "big picture" science that interfaces with and extends narrower disciplines such as economics, agriculture, engineering, history, and sociology.

This is a fun picture. There is room for only one gull on top of each of these tall poles in Bodega Bay, California, and they constantly try to displace one another, emitting raucous cries. One or more species of "seagulls," as they are known to most people, are a nearly

gull musical chairs, CALIFORNIA

universal sight along the shores of seas and large inland lakes. Gulls are opportunistic scavengers, hunters and gatherers like early humans. When fishing boats raise their nets, gulls appear from seemingly nowhere to scramble for anything that falls or is tossed overboard. In the days of open garbage dumps, gulls would clean up anything edible.

In Salt Lake City, Utah, stands a statue of a gull, erected in appreciation for the occasion when gulls saved the early settlers' crops from a horde of migratory locusts. Large flocks of gulls came off the Great Salt Lake and quickly ate up all the locusts before they could consume the crops.

On the west coast of Florida, gulls have learned how to steal fish from pelicans. When the pelican comes up from a dive (see page 69), a gull perches on top of the pelican's head and reaches down to swipe a fish from the pouch as the pelican tilts its head back to swallow the catch. Strangely enough, the pelican makes no attempt to shake off the gull, apparently willing to share its catch with a less expert fisherman.

OPPOSITE: MUSICAL CHAIRS–GULLS AT BODEGA BAY, CALIFORNIA (1969)

Sapelo Island, one of the larger barrier islands on the coast of Georgia, was acquired by the state from the late R. J. Reynolds, tobacco millionaire. The island is currently managed as a marine sanctuary and as an educational, conference, and research center with limited access for the general public for managed deer hunts and guided tours. Martha's pictures of Sapelo from the marsh side are on page 89. These are views of the fore dunes (those closer to the water) on the beach side. The beach and dunes on Sapelo remain natural, with no development other than a boardwalk and an observation shelter.

sapelo DUNES

During the fifty years that the University of Georgia Marine Institute has operated on Sapelo, the beach and dunes have accreted (extended) at the south end of the island and eroded (moved inward) at the north end. Many other barrier islands are also slowly shifting south. Dunes with plenty of vegetation help stabilize, or at least slow changes in, seashores, but not in the way most people think. Sand dunes do not block storm tides, as would a concrete wall, but as the storm water moves over the dune, the enormous energy of the storm is dissipated over a wide area; as the water recedes back over the beach, sand from the dunes replaces the sand washed out by the storm, naturally replenishing the beach.

When sea walls, intended to protect houses, are built too close to the shore, two bad things happen: the energy of the storm tides is reflected by the wall back onto the beach, washing sand back into the ocean; and the source of sand for natural replacement is cut off by the wall. Accordingly, where there are walls, the beach becomes steeper and steeper until there is no beach at normal daily high tides. The house may be saved for the time being, but at a cost of losing the beach, which can be restored only by expensive artificial replenishment. To replace this natural and free process by pumping sand from the depths back onto the beach costs about one million dollars a mile and has to be repeated every few years.

The beach-dune story is a good example of why we need to consider every landscape as an ecosystem, integrated parts that function as a whole. Dunes and beaches function as a system; you cannot deal with only one aspect of this system, as anything you do to one affects the other. A better plan to protect houses is to build them back at the land edge of the dunes and put them on tall pillars. This not only improves the view but also allows the storm water to flow freely by without damage to the house. This mode of construction is now required in the newer seashore developments. It pays to design with natural forces, not against them.

opposite above: Sapelo Dunes (1969)
opposite below: Sapelo Beach (1986)

Lighthouses and bridges appear in many of Martha's paintings; here is one with both, at Jupiter on the Gold Coast of Florida, just north of Palm Beach. This was painted in 1970, the year of the first Earth Day and the beginning of a worldwide concern for the environment.

jupiter lighthouse & bridge, FLORIDA

In the 1960s, astronauts had taken pictures of the earth from space and from the moon that showed how small, fragile, and alone our planet was, suspended in hostile space. It was the first time in human history that we could see the earth as a whole. Seen from space, the east coast of Florida is indeed "golden" in that at night, it is a continuous strip of lights. Wealth and overdevelopment are rapidly covering the natural environment's beauty with concrete and sprawl.

At Jupiter, there is a very exclusive club, whose members are CEOs of many of the largest corporations in the United States. I was invited to speak at this club's annual banquet in April 1970, and I assumed they thought that they should hear something about how industry might help the environment. Banquet speeches are always a tough assignment since usually most of the audience has had too much to eat and drink and expects jokes and entertainment rather than anything of substance. All I remember from that night was that something I said so agitated an executive sitting in the front row that he fell off his chair.

Jupiter Lighthouse, Florida
(1970)

Rocky seacoasts with surf dashing against rocks and sending up plumes of spray always seemed to inspire Martha to paint. The sheer energy of moving water is certainly inspiring and the nearest thing to perpetual motion that we have. Seals, sea otters, and seabirds are

rocky coasts of OREGON & CALIFORNIA

added attractions of rocky coasts. In all, Martha painted more than sixty rocky coast scenes on the West Coast, Maine, New Brunswick, northern France, Japan, and Portugal. The two shown here were among the first of her small, "on the spot" sketches of rocky shores. For another, see page 65.

Beaches on the rocky Big Sur coast of California are restricted to small coves, as shown in the lower picture. Many of these beaches are seasonal in that they appear and disappear in an annual cycle. During the rainy season, in winter and spring, sand washed down streams and rivers builds up in the coves; during the dry season, in late summer and fall, most if not all of the sand may wash out. The famous bridge on Route 1 at Big Sur is shown in the pen-and-ink sketch above.

Of course, all beaches come and go, but usually over longer periods of time. On low-lying coasts, humans often hasten beach washout by building sea walls, which can increase beach erosion in two ways. For more on why it is better to design with natural forces than confront them, see page 58.

OPPOSITE ABOVE: Seal Cove Rocks, Oregon (1970)

OPPOSITE BELOW: San Pedro Point, Montara Beach, California (1970)

M. ODUM

After a seminar visit with students and professors at Bowdoin College in Maine, we rented a cabin on nearby Pemaquid Point, one of many narrow rocky peninsulas that jut into the sea on the coast north of Portland. While I worked on a revision of my textbook, *Basic Ecology*,

pemaquid point, MAINE

Martha painted. One day, she spent an entire tidal cycle sitting on a folding chair at the very end of the point, painting a series of pictures. She started with low tide, when the water was relatively calm, as in the picture reproduced here, and ended at high tide, when the waves crashing against the rocks sent up spray ten feet high. She called this series her "Maine Symphony."

Later, in a one-woman show in Athens, she exhibited the "Maine Symphony" along with other paintings of both Atlantic and Pacific rocky coasts.

PEMAQUID POINT, MAINE
(N.D.)

Mel Warnick was one of Martha's close friends when Martha was actively involved in building a house for the Delta Gamma sorority at the University of Georgia, and Mel was the chapter's first house mother. Mel owned a small island with a cabin on it in Lake of the Woods, Canada, which we visited in August 1971.

mel's island, LAKE of the WOODS, CANADA

Northern lakes, in contrast to energetic coastlines, have a tranquil beauty, especially when autumn colors begin to develop, in late summer. The wild cries of loons at night add to the mystique of northern lakes. Sounds associated with particular landscapes are part of the essence of place as much as sights.

When we visited Mel's island, she was having trouble with beavers that insisted on cutting down trees close to the cabin that she did not want cut down. In general, native wildlife species play important roles in natural ecosystems, and many species receive some kind of protection or management from humans. Beavers diversify the landscape by building ponds and creating openings in the forest that become the homes of numerous species (fish, frogs, water birds, dragonflies) that are not present in unbroken forest. When they become too numerous or invade too closely into our living space, beavers can become pests. We should remember that the beaver was very important in the early economy of this country, since beaver pelts were the currency of trade among early settlers and between settlers and the Indians.

OPPOSITE: VIEW FROM MEL'S ISLAND, LAKE OF THE WOODS, CANADA (1971)

The pelican is a bird whose "bill can hold more than his bellican," or so say postcards from Louisiana, where it is the state bird. Pelicans fish by diving headfirst into the water. Small flocks of brown pelicans cruising low over the surf are a common sight

pelicans of St. Croix

along the southeastern Atlantic and Gulf coasts and south into the Caribbean and northern South America. White pelicans are found in the western United States, and a few migrate east for the winter.

We almost lost the brown pelican in the 1960s due to the pesticide DDT, which in birds causes shell thinning to the point that no eggs can survive to hatching. Fish-eating birds such as pelicans and ospreys are especially vulnerable to this kind of pollution because of what we call biological or food chain magnification. This occurs when very small amounts of poison in the water are concentrated, or magnified, with each step up in the food chain, i.e., the progression from microorganisms to small invertebrates to small fish to large fish to fish-eating birds. The top predators thus get what otherwise would not be a lethal dose. A decline in these predators provides a warning to humans, who are also top predators.

The pelican completely disappeared from the Gulf coast, presumably because of the large amounts of pesticides coming down the Mississippi from the grain fields of the Midwest. After DDT and other chlorinated pesticides were banned in the 1980s, the pelicans and ospreys made a comeback.

The exuberance of pelicans, who really seem to enjoy flying, diving, and swimming, is captured in Martha's watercolor sketch, which she painted while we were visiting St. Croix in the U.S. Virgin Islands.

Brown Pelicans, Tague Bay,
St. Croix (1972)

The Grand Canyon is a spectacular and colorful natural wonder that is visited by millions of people not only from the United States, but from all over the world. Even more colorful, but much less well known, are the vast canyon lands of southern Utah that lie

view from dead horse point, UTAH

upstream from the Grand Canyon and the Hoover Dam. The high point known locally as Dead Horse Point near Moab is a good place for a panoramic view of this kaleidoscopic landscape where colors, changing with the shifting of sun and shadow, create a real challenge for any painter.

There are very few roads and almost no towns in this rugged country, but lots of wildlife. It's a great place for backpacking and boating up and down the numerous canyons cut by the streams that flow into the Colorado River. Let us hope that one hundred years from now this country will still be wild.

Dead Horse Point,
Near Moab, Utah (1972)

monument VALLEY

Earlier in this book, I mentioned that the Monument Valley–Four Corners area of Arizona, New Mexico, Utah, and Colorado was one of Martha's favorite places to paint. These two pictures from 1972 were among the first she painted in the Southwest. For the next twenty years, I rarely missed an opportunity to attend a conference or give a college seminar in this region so that Martha could paint there. The attraction of this landscape to the artist is the variety of shapes and colors of the sandstone "monuments," carved by nature rather than by humans. For other views of this landscape, see pages 77 and 129.

The Four Corners region is the home of the Navajo tribe. We both became very much interested in Navajo and other Southwest Indian art (sand paintings, blankets, pottery) and culture, as it is closely intertwined with the environment. One of our favorite authors is Tony Hillerman, whose Navajo mystery novels contain vivid descriptions of the mountains, deserts, and weather events as viewed by the Indians. Also, one of Martha's favorite artists was Georgia O'Keeffe, who spent half of her life painting on her ranch in northern New Mexico.

Incidentally, there is an amazing diversity of cultures in northern New Mexico. On the one hand, numerous Indian subtribes each celebrate a different deity with traditional dances; at the other extreme is the Los Alamos Atomic Energy Laboratory, where the atomic bomb was born. On top of a mountain is the Santa Fe Institute, a think tank where scientists and philosophers come together to try to understand and deal with complex systems. One person there told me, half in jest, that they are trying to find out where science (the explainable) ends and religion (the unexplainable) begins.

OPPOSITE ABOVE: Monument Valley, Arizona, Navajo Tribal Headquarters (1972)

OPPOSITE BELOW: Valley of the Gods, Mexican Hat, Utah (1972)

M. ODUM

In the Utah corner of Monument Valley is a place called Goulding's Lodge and Trading Post, which has an adjoining motel. Every room in the motel has a balcony with a panoramic view of the vast desert landscape with its diversity of red sandstone "monuments."

monument VALLEY at SUNSET

One year, we were here on Martha's birthday, in May. She spent most of the day on the balcony, painting a series of pictures from dawn to dark to capture the scene under different light conditions and cloud formations. The view reproduced here is at sunset, when the light is beginning to fade, and you can almost see the ghosts of the Indians in the shadows.

Monument Valley, Utah, Sunset as Seen from Goulding's Trading Post (1972)

Both flowers and canals are characteristic features of the Dutch landscape. Many of the flowers, bulbs, and seeds you buy from U.S. florists have been shipped by air freight from Holland and constitute a major export for the country. As noted in the commentary on page 20, one

flower market HOLLAND

can make a good living there from a small plot of land if one grows flowers rather than field crops (corn, soybeans, wheat), which require large tracts of farmland and expensive chemicals and machinery to turn a profit.

While I was attending the first International Congress of Ecology in The Hague, in August 1974, Martha rented a bicycle and pedaled around the countryside, stopping here and there to sketch or visit art galleries and museums. The country has a network of bicycle paths, and Dutch motorists look out for bicycles more than American drivers do. In the sketch opposite, flowers at a flower market are reflected in the water of a canal in the small town of Delft.

OPPOSITE: FLOWER MARKET, DELFT, HOLLAND (1974)

After my brother and I had received the international Prize of the Institut de la Vie in Paris in 1975, Martha and I rented a car and went on a painting tour of the Brittany coast of France, west of Normandy Beach, the site of the Allied invasion in World War II. Much of this coast

the brittany coast of FRANCE

reminds one of California's Big Sur coast, rocky with high cliffs, but also possessing sandy beaches and boat harbor coves, a very scenic mix. The pen-and-ink sketch above is of another point on the Brittany Coast.

Tides run high along this section of the coast, and one of the world's few operational tidal power plants, which we visited, is located here. The generators hang from a low bridge across a narrow channel, and their blades turn on both the flood and ebb of the tide as the water flows strongly back and forth through the channel. Tidal power cannot contribute much to our electrical needs, but sedentary marine organisms such as oysters, clams, mussels, and tube worms benefit from tides, which bring in food and take away wastes. As a result, density and diversity of marine life is very high in tidal communities, some of which are the nearest thing in nature to human cities, requiring large inputs of energy and food and large export of wastes. See page 10 for the concept of a city's "ecological footprint."

Cap d'Erquy, Coast of Brittany, France (1975)

To Martha, an airplane was a means to see the big picture of earth and clouds as well as a means of transportation. Whenever possible, we would book a window seat back of the wing for her, where there was an unobstructed view. She would whip out her sketch

views from AIRPLANES

pad and paint box from her purse and capture it. I have selected two samples from her collection of views from airplanes, one of cloud formations and one of the land-use pattern of an English landscape as seen from a low altitude.

Where humans are numerous and active, the landscape gets to look like a checkerboard, with patches and strips of different habitats and land uses, like the neat squares of cropland seen in the picture of the English countryside. There is increasing concern worldwide about excessive fragmentation of the landscape. For a discussion of the establishment of natural buffers such as greenbelts to preserve biodiversity and air and water quality, see page 6.

OPPOSITE ABOVE: Clouds from Plane Window (1974)

OPPOSITE BELOW: English Countryside as Seen from Plane (1974)

M. ODUM

The Snake River runs west and north almost the whole breadth and length of the state of Idaho before turning west into Washington to join the Columbia River. Although there are four dams on the Snake in Idaho, there remain long stretches that are free-flowing and relatively pristine.

snake river CANYON

The Nature Conservancy has acquired a section where the river "snakes" (i.e., zigzags) through a deep canyon for permanent preservation as a natural area. When I was a member of the Conservancy's board of trustees, Martha and I joined the trustees for a two-day float down this canyon in May 1975.

The sanctuary area is of special interest to conservationists because it provides nesting sites for large birds of prey, including golden eagles, prairie falcons, and several species of large broad-winged hawk. These birds find safe nesting places on ledges on the canyon walls. For the most part, raptors don't get their food from the canyon itself, but from the farm and ranch land that borders the canyon, where ground squirrels and rabbits live. These large, graceful birds fly straight up and over the canyon wall to hunt and then return in a nosedive with prey in their talons, quite a spectacular sight.

The two watercolors opposite illustrate our overnight camp site. One can see rafts and a patrol boat in the upper picture and a tent and a group of trustees around a campfire in the lower picture. We opted to spend the night in sleeping bags in the open rather than in a tent. Waking up to the sound of running water and the songs of the canyon wren was worth sleeping on the ground on a cold night. Wrens are very small birds, but all species (including our eastern Carolina and house wrens) have songs that are surprisingly loud. The canyon wren's song, consisting of a string of bell-like notes, reverberates between the canyon walls and is one of the great sounds of the western canyons, including of course, the Grand Canyon.

opposite above: Snake River Canyon, Our Raft in the Foreground (1975)

opposite below: Snake River Float Trip with the Nature Conservancy Board Breaking Camp in the Morning (1975)

JAWS

Our trip to the Canadian maritime province of New Brunswick in July 1975 was a combined mission of art and ecology, an occasion for both of us to display our wares, as it were. My former student "Buster" Welch arranged for me to give a lecture at the biological station where he

fireweed & fishtraps, NEW BRUNSWICK

researched intertidal life and for Martha to have a one-woman show of her water-colors at a gallery in St. Andrews. After the lecture and show, we explored the Bay of Fundy coast, providing Martha with another opportunity to add to her collection of seacoast paintings.

Marine fish are a very important resource in the maritime provinces. The remains of a fish trap seen in Martha's painting is an example of an early means of fish harvesting no longer in general use. The trap consists of a maze of fences into which fish wander and are unable to find their way out.

The fireweed on shore is a tall plant (belonging to the evening primrose family) with a spire of pinkish-red or rose-magenta flowers. It grows in openings and clearings in northern forests ranging across Canada, south along the mountains to Georgia, and in the west, south to California. It is one of the first plants to appear after a forest fire, often forming dense stands, converting the scorched earth into a beautiful landscape.

Fireweed also grows in Great Britain on disturbed places along railroads and roads. During World War II, it appeared in abundance in bombed and burned out areas of London.

Fireweed and Fish Traps,
New Brunswick, Canada (1975)

The coast of New Brunswick is rugged, like Maine, but colder and with fewer resident humans and more nesting seabirds. Fishing is a way of life for both humans and birds since the offshore waters are very productive, despite the low temperatures. In fact, the low temperatures actually

oven head point, NEW BRUNSWICK

reduce the amount of zooplankton, or microbial life, in the water. Zooplankton consume phytoplankton (photosynthetic green floating life), which live in the upper sunlit waters, but in this scenario, more phytoplankton reaches bottom-feeding fish such as cod.

Most of the paintings Martha did on this trip featured the sea, but as a change of pace, she decided to do this one of the land. On land, the forests are dominated by spruce and fir. These two northern conifers look the same from a distance, but if you want to know the difference, grab hold of a branch. If it feels prickly and has stiff needles, then it's a spruce; if softer, it's a fir. Also, spruce cones generally hang downward, while fir cones sit upright on the branches.

Near Oven Head,
New Brunswick (1975)

The Georgia and South Carolina coastal marshes and barrier islands (known as the Sea Islands to most people) are something special. Views across this landscape of endlessly intricate patterns of grass beds, creeks, lagoons, and forested islands vary with time of day, weather, and season. As shown by two of Martha's paintings of Sapelo Island from the marsh side, the view can vary from green and blue patterns in bright summer sunlight to golden hues in the fall.

georgia salt marshes at SAPELO

The Georgia and South Carolina marshes are more extensive and better developed than those either north or south of the two states because of their location in the Southeastern Bight, the name given to the indentation of the coastline between Myrtle Beach, South Carolina, and Jacksonville, Florida. Here, the tidal amplitude (the difference between mean tide and high tide) is six feet or more, greater than either to the north or south, due to the longer "fetch" from the offshore gravitational dome resulting from the pull of the moon and sun (spring tides occur when both pull together). This greater water flow irrigates and increases the productivity of the marsh-estuary system, leading to taller grass and more nursery grounds for shrimp and fish.

Our research at the University of Georgia's Marine Institute on Sapelo, which began in 1954, played a major role in the passage of statewide coastal marsh protective legislation in 1970. In addition to our work demonstrating the value of the marshes, Sunday supplements, editorials, cartoons, national magazine publicity, constructive student activism fact sheets widely distributed to schools and citizens, unified support from conservation organizations, bumper stickers and buttons, and especially letters to legislators all played a part in convincing the Georgia legislature that the half-million acres of marsh-estuary should be protected from drainage and development. Neither the conservative legislature nor the governor was in favor of the bill at that time, but the overwhelming public support resulted in near-unanimous passage of the protective legislation. The situation illustrated that, in general, politicians do not lead; they follow. When a majority of people indicate that they want something done, it gets done!

OPPOSITE ABOVE: Sapelo, Marsh-side, Low Tide (ca. 1976)

OPPOSITE BELOW: Sapelo Marsh (Fall) (n.d.)

M. ODUM

As noted in the commentary on page 44, Martha loved boats and was an accomplished sailor, so boats of all kinds show up in her paintings. In the upper picture, two shrimp boats are moored at the dock of the small town of Valona, Georgia, located on the

shrimp BOATS

mainland edge of the wide salt marsh estuary; in the distance is Sapelo Island, where we spent a lot of time at the University of Georgia's Marine Institute.

In the lower picture are several shrimp boats with their nets hanging up to dry at Ft. Myers, on the Gulf coast of Florida. Some of these boats are larger than the ones at Valona and are capable of fishing not only the Florida Gulf waters but also far to the south, to the Yucatán and beyond.

Shrimp are an annual crop. Once the adults leave the inshore nursery grounds, they spend the rest of their short life offshore. In theory, shrimping is sustainable so long as harvesting is restricted to offshore waters and dragging is not permitted in the inshore sounds and marsh-estuaries. A shrimp boat with nets down (dragging) off Sapelo beach is shown in the pen-and-ink sketch above.

opposite above: Valona, Georgia (Sapelo Island in Distance) (1978)

opposite below: Shrimp Boats, Fort Myers, FL (n.d.)

Among butterflies, there are a number of mimics, that is, species that benefit in terms of natural selection by having the same coloration and general appearance as another, unrelated species. The viceroy mimics the monarch, which is famous for its long-distance migrations

viceroy BUTTERFLIES

to wintering grounds in the mountains of central Mexico. Monarchs are very distasteful to eat; their body contains a powerful emetic (an agent causing vomiting) so predators learn quickly not to eat them. The viceroy does not have this repellent poison, but avoids being eaten because would-be predators mistake it for the monarch. The two species are so similar in their orange-and-black coloration that one has to look very closely to tell them apart. For example, the monarch has two rows of tiny white spots in its black wing borders, while the viceroy has only one row.

Viceroys do not migrate as far as monarchs, but small flocks move south in the fall along the southeastern coast of the U.S. Martha's picture shows viceroys seeking nectar from a late-blooming coastal shrub on the Georgia coast in September. Also, in the bottom of the painting is a sphinx moth, a species with a large body and small wings. Unlike the butterfly, the sphinx flies with a rapid wingbeat, like a bee or hummingbird, demonstrating that there is more than one way for moths and butterflies to fly.

Viceroy Butterflies,
Valona, GA (1978)

yellow bark acacias, KENYA

Within sight of the skyscrapers of Nairobi, the capital of and largest city in Kenya, is a remarkable game park with free-roaming antelopes, giraffes, lions, and many other large African animals. We were on tour in this park when our safari bus got stuck (see the pen-and-ink sketch above). In the thirty minutes or so required to get the vehicle out of the mud, Martha painted this watercolor of the picturesque flat-topped acacia trees that are so characteristic of the East African savannas (grasslands with scattered trees or clumps of trees).

Since plants cannot run away and hide when attacked by insects or grazing animals, they have to defend themselves by producing thorns, tough coatings on leaves, and especially "anti-herbivore" chemicals that repel the would-be eaters. Some species of acacia have developed another defense strategy: they essentially hire an army of ants to defend them. They do this by secreting sugars all along their stems and leaves and providing small cavities in their thorns and branches that the ants use for nests. As the ants cruise over the tree lapping up the food, they kill or scare away herbivores. Thus, the plant provides room and board in return for protection.

Most people do not realize how much cooperation between unrelated species for mutual benefit there is in nature; it's not all cutthroat competition or "dog-eat-dog" by any means. The message for humans is that the time has come to promote a more harmonious coexistence between nature and humankind for the good of all.

Yellow Bark Acacia Trees Near
Nairobi, Kenya (1979)

The bar at the lodge at Samburu, Kenya, has a terrace that looks out on a large mud bar in a small river to which crocodiles and large Malibu storks come to feed on meat and fish scraps from the kitchen. Thus, the lodge's "watering hole" is appropriately called the Crocodile Bar.

the crocodile bar at SAMBURU, KENYA

Also, fresh meat is suspended from a tree on the opposite bank to attract leopards (see sketch below). You do not have to drink a lot to see scary creatures; they are there in real life.

Storks, incidentally, are an interesting group of birds, including the legendary European stork that nests on chimney tops and supposedly brings babies. Storks are easily distinguished from herons and other wading birds by their large, heavy bills that point downward. There are a number of different species in Africa, the Malibu being the largest; only one, the wood stork, is native to North America, being found in Florida and adjacent southeastern states.

Crocodile Bar,
Samburu, Kenya (1979)

tree tops lodge, KENYA, & BABOONS

The Tree Tops Lodge is built on top of tall poles in a wildlife preserve in Kenya. The lobby is on the flat roof, where guests can view and photograph the animals, such as the African buffalo and the warthogs seen in Martha's sketch. There are chairs and tables on the roof, and snacks and drinks are served. One does have to watch out for the baboons, which can get on the roof by climbing up the side of the building or jumping from nearby trees. The half-tame animals are very quick to snatch food from the tables or even out of your hands, so the waiters carry big sticks to keep them at bay.

Late in the afternoon, when Martha was painting the scene with the buffalo and warthogs, a baboon suddenly jumped on her shoulder and made a grab for her paint box, apparently mistaking it for food (see pen-and-ink sketch above). Realizing his mistake, he quickly jumped off without doing any damage, but he did leave some distinct paw prints on her jacket, which Martha would show friends at home to back up her story.

The next day Martha did a portrait of a well-behaved mother baboon and her baby, resting quietly in a nearby tree. Baboons are crafty opportunists, like many people we know.

ABOVE: AFRICAN BUFFALO AND WARTHOGS AS SEEN FROM TREE TOPS (1979)

OPPOSITE: OLIVE BABOON, MOTHER PROTECTING BABY IN MISTY RAIN, TREETOPS, KENYA (1979)

MARTHA ODUM

The year I was a visiting professor for one week at Colorado State University in Fort Collins provided Martha with opportunities to explore and paint two of her favorite rocky rivers, the Cache la Poudre and the St. Vrain, which flow out of steep-walled canyons onto

cache la poudre river, COLORADO

the plains. The sheer walls and huge boulders in both river canyons result in spectacular torrents of water.

A couple of years after the picture shown here was painted, an August thunderstorm got "stuck" upstream. Normally, thunderstorms move along so the heaviest rains do not last long in any one place. But in this case, the storm stalled, creating a cloudburst resulting in a flash flood that washed out a good part of the highway that parallels the river in the narrow canyon. Some cars were washed downstream, but as I recall, no one was killed. Flash floods do, however, kill people, so I was glad Martha was not up there at the time.

OPPOSITE: CACHE LA POUDRE RIVER, COLORADO (1979)

The ski season may be over by the middle of May, but patches of snow often linger, and many ski lodges remain open. During the off-season, the lodges book conferences for people to come discuss business, science, or whatever and, of course, enjoy the views.

crested butte ski lodge, COLORADO

While I was attending an ecological meeting at the Crested Butte Lodge, located in the Rocky Mountains, Martha was out painting, as would be expected. Her sketches are much better than any diary or long-forgotten conference program folders to remind us that we were there in 1979.

The essence of this place is that the winter-summer transitional landscape is much more interesting than an all white winter or an all green summer one. Ecologically, this picture reminds us that snowmelt is a major source of water for cities like Denver. When the winter snowpack is light, water shortages on the plains below follow.

Mt. Crested Butte, Ski Area (1979)

The Rocky Mountains run through the middle of Colorado. This view is from the back side, or from west to east, as seen from the town of Gunnison. As anyone who watches The Weather Channel knows, weather fronts or systems in the U.S. move from west to east and from south

view from gunnison, COLORADO

to north; most precipitation comes from the Pacific and Atlantic Oceans and from the Gulf of Mexico. As the moisture-laden clouds move east from the Pacific, much of the water falls on the western side of the Sierra Nevada Range, creating dry areas known as "rain shadows" on the east side of the high mountains.

Death Valley is an extreme example of a rain shadow. Western Colorado is somewhat affected by the Sierra Nevada rain shadow, so low elevations are semi-arid, as seen in the foreground of Martha's painting. Then the Rocky Mountains create a second rain shadow in eastern Colorado, where the landscape is dominated by semi-arid grasslands, ranches, and irrigated croplands. Colorado provides a wide range of climates, and Martha's picture presents the essence of place for the whole state.

OPPOSITE: EDGE OF GUNNISON, COLORADO (1979)

The eastern Appalachian Mountains have two great trails that run along the backbone of the high country. One, the Appalachian Trail, extending from Maine to Georgia, is for hikers. The other, the Blue Ridge Parkway, is for automobiles. Martha's October picture in Virginia

the blue ridge PARKWAY

shows the first autumn colors on the parkway and the blue mountains in the background.

Many people drive the length of the Blue Ridge Parkway, but very few walk the entire Appalachian Trail. Our Athenian friend Fred Birchmore is one who has gone the distance. Fred is a small man, all muscle and sinew, and has remained physically active well into his eighties. He is also a good naturalist who enjoys the plants, birds, and other wildlife encountered on the trip. He prefers tennis shoes to hiking boots for long hikes and wore out three pairs on his jaunt from Georgia to Maine.

OPPOSITE: BLUE RIDGE PARKWAY (1979)

It has not been long since wetlands, or swamps, were not only considered worthless, but also regarded as breeding grounds for mosquitoes, diseases, and poisonous snakes; therefore, people thought they should be drained as soon as possible. Then, as scientists began to

freshwater WETLANDS

document that wetlands were home to valuable wildlife, often very productive, and a source of valuable ecosystem services such as pollution abatement, public opinion switched dramatically. Today, not only is wetland preservation promoted, but so is wetland restoration. In fact, restoring drained wetlands and constructing new ones to deal with mine wastes and other pollutants is part of the new field of environmental engineering that is providing business opportunities and jobs for ecologists. Our research at the University of Georgia's Marine Institute on Sapelo Island has played a role in this attitude change about wetlands, as noted on page 88.

The wetland in Martha's painting is near Atlanta's Chattahoochee Nature Center, where I was taking part in a program on wetland values while Martha sketched. It serves as an outdoor classroom for environmental education programs organized by schools and nature centers in the Atlanta area.

OPPOSITE: WETLAND NEAR CHATTAHOOCHEE NATURE CENTER (1979)

the marshes of GLYNN

Sunsets can be spectacular over the coastal marshes. These marshes at Saint Simons Island in Glynn County, Georgia, are poet Sidney Lanier's "Marshes of Glynn." Here are some excerpts from that poem written more than a century ago, in 1878:

Ye marshes, how candid and simple and nothing-withholding and free
Ye publish yourselves to the sky and offer yourselves to the sea!
Tolerant plains, that suffer the sea and the rains and the sun,
Ye spread and span like the catholic man who hath mightily won
God out of knowledge and good out of infinite pain
And sight out of blindness and purity out of a stain.
As the marsh-hen secretly builds on the watery sod,
Behold I will build me a nest on the greatness of God:
I will fly in the greatness of God as the marsh-hen flies
In the freedom that fills all the space 'twixt the marsh and the skies:
A league and a league of marsh-grass, waist-high, broad in the blade,
Green, and all of a height, and unflecked with a light or a shade,
Stretch leisurely off, in a pleasant plain,
To the terminal blue of the main.

Oh, what is abroad in the marsh and the terminal sea?
Somehow my soul seems suddenly free
The creeks overflow: a thousand rivulets run
'Twixt the roots of the sod; the blades of the marsh-grass stir;
Passeth a hurrying sound of wings that westward whirr;
Passeth, and all is still; and the currents cease to run;
And the sea and the marsh are one.
Till his waters have flooded the uttermost creeks and the low-lying lanes,
And the marsh is meshed with a million veins,
That like as with rosy and silvery essences flow
In the rose-and-silver evening glow.
And I would know what swimmeth below with the tide comes in
On the length and the breadth of the marvellous marshes of Glynn.

As a result of research on Georgia's salt marsh estuaries at Sapelo Island, which began in 1954, we now know quite a bit about "what is abroad in the marsh and the terminal sea" and "what swimmeth below" when the tide comes in.

St. Simons Island, GA
(1980)

To those who live in the western United States, water is such a valuable and scarce commodity that it is considered a terrible waste to let any of it to flow out to sea. Every effort is made to slow down the flow with dams and to withdraw as much as possible before it reaches the ocean. In

american river, CALIFORNIA

flood-prone regions of the East, the mind-set is, of course, just the opposite: the more water that flows out to sea the better.

Even small rivers in California may have as many as four dams, so free-flowing stretches, such as this one on the American River, are few. As noted previously in this book, Martha had a special talent and fondness for depicting water flowing over rocks.

We humans tend to go too far with good things; we do not always know when bigger is no longer better. In general, dams and impoundments are good things in that they provide water storage, water power, and recreation, especially in regions that lack natural lakes. In recent years, though, we have seen that it is no longer desirable to dam up every stream and that we now have too many dams. As a result, unnecessary dams, especially those on small rivers, are now being demolished.

One of my favorite environmental journalists is John McPhee, who writes on humans' often misguided attempts to control nature. In the September 27, 1999, issue of the *New Yorker*, he wrote an article entitled "Farewell to the Nineteenth Century: The Breaching of the Edwards Dam." The Edwards Dam, on the Kennebec River in Maine, was built in the early 1800s, when it powered sawmills, gristmills, and machine shops, but its economic value had deteriorated to virtually nothing. Breaching this dam restored salmon runs and washed out accumulated sediments and pollution. Rapids and streamside, or riparian vegetation, like those in Martha's painting, have replaced impounded still water on the Kennebec.

American River,
Near Folsom, California (1980)

Although Martha and I prefer small towns and the countryside to large cities, Chicago has been important in our lives. Martha grew up in the north shore suburb of Wilmette, and her first job was in Chicago, designing wallpaper. We also spent our honeymoon there, and we both

view from the drake hotel, CHICAGO

obtained our terminal degrees at the University of Illinois, Urbana-Champaign, a hundred or so miles to the south.

The Drake is one of Chicago's grand old hotels; most rooms have a great view of Michigan Avenue and the lakefront park. Martha's sketch was done on Sunday morning, when there was little traffic. Chicago is an exciting and diverse city where one can find most anything, including parrots trying to establish themselves as wild, urban birds.

There is only one parrot native to North America, the Carolina parakeet (small parrots are called parakeets), which is now extinct. Several species of South American parrots, however, brought into the U.S. as caged birds, have escaped or been deliberately released and have established themselves as wild birds in Florida. One species, the monk parakeet, has not only become established in Florida, but has also appeared in northern cities, including Chicago and New York. It is about the size of a mourning dove, green topside with a gray head that resembles a monk's hood, hence the name. Unusual among parrots, the monk builds large stick nests that may house more than one pair on building ledges or in the tops of tall trees. They apparently survive the winter by visiting people's bird feeders and eating sunflower seeds.

View from the Drake Hotel,
Chicago (1980)

Nova Scotia juts out into the cold North Atlantic, and so is often surrounded by clouds, fog, and wind. Martha's painting captures the essence of place of this chilly landscape. On the more positive side, the sea is fertile, there are a lot of fish, and spectacular seabird nesting

whale cove, NOVA SCOTIA

colonies fill some of the cliffs from top to bottom.

One of my former students, George Childs, after retiring from academe, decided he wanted to try farming in an ecologically sustainable manner. He bought a farm in Nova Scotia, not the best place in the world for farming, but he told us it was the only large tract he could afford. He did, however, find a "crop" that was both profitable and good for the environment: draft horses.

The Childses built an octagonal house heated by a large wood-burning furnace. They told us that it was easier to heat, and the winter winds did not whistle and howl as they do around a square house because of the absence of sharp corners.

Near Whale Cove, Cape Breton,
Nova Scotia (1981)

City Hall sits on top of a hill amid tall trees. The architecture is unusual in that the building has a clock tower, two domes, one on top of the other, and a spread eagle on top. When our son was about two and a half years old, he pointed to the eagle as we were driving

athens, georgia, CITY HALL

by and said, "Look, momma, bird stuck!"

City Hall in Athens is a very active place, with people coming and going, attending civic meetings or dealing with traffic fines or other transgressions in city court. The old wooden floors and staircases are worn down as a result of all the human traffic over the years.

Out front is the famous double-barreled cannon of Civil War times. It never worked because the two balls, connected by a chain, failed to emerge at the same time; the steel curtain that was supposed to mow down the enemy completely failed, but did make it into "Ripley's Believe It or Not."

Athens City Hall Eagle
(1983)

The University of Georgia was the first chartered state university in the U.S. and celebrated its bicentennial in 1984. The building to the left in Martha's painting of the old campus is Old College, one of the first two buildings that originally housed both students and teachers. The

old campus, UNIVERSITY of GEORGIA

building to the right is the chapel, where lectures and musical events continue to be held. The first faculty were Yale graduates, and the first buildings were copies of Yale buildings. The UGA football mascot, the Georgia Bulldog Uga, is the prodigal son, as it were, of the Yale Bulldog.

The earliest pictures of the university show a small cluster of buildings in a completely open landscape, with no trees of any kind. Pioneers first clear the land, then build a house, plant crops, or whatever. Unfortunately, many developers continue to want to operate that way. On the old campus today, you can hardly see the buildings for the forest. Trees are appropriate for college campuses. Not only does the shade moderate temperatures, but it also creates a tranquility that is conducive to scholarship.

Old Campus, University of Georgia,
Late Spring (n.d.)

Waterfalls are musical landscapes, not only beautiful to observe but also pleasant to hear. The gentle sounds of ocean surf or waterfalls encourage sleep; "white noise" is often recorded for that purpose. In forested regions, waterfalls provide open corridors where sunlight can reach ground level. Accordingly, flowering shrubs such as the purple rhododendron in Martha's picture grow along the banks. For a close-up of this flower, see the next entry.

north georgia WATERFALL

The downhill flow of water creates an indirect form of solar energy. One third of the solar energy that reaches the earth is dissipated in driving the water or hydrological cycle, which involves the evaporation and desalination of water from the oceans, the formation of clouds, rainfall, and the downhill flow back to the sea. This cycle is one of the most important life-supporting services provided by nature. Our water bills include only the cost of pumping and purifying. In today's marketplace, ecosystem services such as water recycling are considered to be free externalities; they have no value until they become scarce. Since freshwater is probably our most endangered resource worldwide, it is urgent that we recognize that water is a very valuable resource and "internalize" this value into the market system, which means greatly improving water conservation and paying more for it.

Mountain Waterfall, Georgia (n.d.)

In general, Martha painted flowers as part of landscapes, but from time to time, she painted individual flower species in her indoor studio, as with this purple rhododendron.

The plant genus Rhododendron (scientific names are usually printed in

purple RHODODENDRON

italics) includes the deciduous azaleas, numerous native species and cultivated varieties, and the evergreen rhododendrons, with three species native to the southern Appalachians and northern Piedmont: a white one, a purple one, and a pink one. The white one (*R. maximum*) forms dense shrub layers in moist mountain forests. The purple one (*R. catabiense*) is also common in southern mountains, especially along streams (see previous entry), and the pink one (*R. minus*) is found on north-facing slopes along streams in the Piedmont. All three are widely planted as ornamental shrubs.

Purple Rhododendron
(n.d.)

After forty-five years or so of Martha tagging along on my conference and seminar trips, I asked her, in 1985, what new place she would like to paint, and it would be my turn to tag along. She picked Portugal, so we went there, rented a car, and drove south from Lisbon to

the cliffs of SOUTHERN PORTUGAL

the southernmost coast, which is the point farthest west in continental Europe. Before Columbus, the Portuguese would look out from high cliffs like those in the picture opposite and wonder what was out there and where the end of the world might be. When ships wrecked on the rocks below, the rescued sailors would be quizzed as to where they had been and what they had seen. It has been human nature through the ages to be curious about the unknown and to want to explore it.

We stayed in small hotels, or *pousodas,* delightful government-run bed-and-breakfasts located in old castles or other historic places. For about a week, Martha painted a large picture every morning while I meditated or looked for birds. After a picnic lunch (see the pen-and-ink sketch above), she would do another picture, followed by a drive around the vicinity to investigate old Roman ruins and other tourist spots. Finally, we would have a big dinner at the inn. It doesn't get any better than that.

Wild Iris–Cabo de São Vicente
Near Sagres, Portugal (1985)

In the desert region of southern Arizona, not many rivers flow all year long, but the one that runs through the Aravaipa Canyon does. The Nature Conservancy has acquired most of this canyon for preservation as a natural area. When we visited this area in October 1993, a former student of mine, Jay Schnell, and his wife, Ginny, were employed to take care of the preserve and operate a guest house for visitors. As it turned out, the pictures Martha painted here were her last before she developed cancer.

the aravaipa canyon, ARIZONA

The essence of this place lies in the unusually colorful canyon walls and in the abundance of bighorn sheep that can be seen on top of the ridges. An unusual Mexican bird, the black hawk, found only in southern Arizona within the U.S., also lives here. Fish are part of its diet, but it does not fish like a fish hawk, or osprey, which hovers over the water and plunges into it from above to grab the fish. Instead, the black hawk stands on the edge of a flat rock in the river where the channel is narrow and reaches out with a long yellow leg to snare a fish as it swims through the shallow rapids. Jay has written the definitive life history of this species in the monographic series *Birds of North America*, published by the American Ornithologists' Union and the Academy of Natural Science of Philadelphia.

Aravaipa Canyon, Arizona
(1993)

Checklist for

Essence *of* Place

1. *Picture Painted from Memory After Ride Through Hills of Wisconsin Near Baraboo in 1938*, 1938
WATERCOLOR
8 1/8 x 12 inches

2. *Our First Home, Rensselaerville, NY*, 1939-40
PEN & INK & WATERCOLOR
8 1/8 x 10 1/8 inches

3. *View from Hotel Trietem, Norway*, 1958
HEAVY PEN & INK & WATERCOLOR
11 x 8 3/8 inches

4. *Millport, Isle of Cumbrae, Scotland, Red Sandstone and Beach*, 1958
WATERCOLOR
8 1/8 x 11 inches

5. *Lake Windermere, Cumbria*, n.d.
WATERCOLOR
8 x 10 5/8 inches

6. *Yakima River, Horse Heaven Hills in the Distance, Eastern Washington*, 1958
HEAVY PEN & INK & WATERCOLOR
8 3/16 x 11 inches

7. *Ruby Beach, Olympic Peninsula, WA*, 1958
HEAVY PEN & INK & WATERCOLOR
8 1/8 x 11 inches

8a. *Taxco, Mexico #1*, 1960
PEN & INK & WATERCOLOR
4 7/8 x 7 inches

8b. *Taxco, Mexico #2*, 1960
PEN & INK & WATERCOLOR

9. *Nagasaki, Japan*, 1962
PEN & INK & WATERCOLOR
8 1/8 x 5 1/2 inches

10. *Shirayama, Japan*, 1962
PEN & INK & WATERCOLOR
5 1/2 x 8 1/8 inches

11. *Shinano Omachi, Japan, View from Hotel Kanden Club*, 1962
PEN & INK & WATERCOLOR
5 1/2 x 8 1/8 inches

12. *Japanese Monkey, Beppu*, 1962
PEN & INK
8 1/8 x 5 1/2 inches

13. *Mt. Fuji, Japan*, 1962
PEN & INK & WATERCOLOR
5 1/2 x 8 1/8 inches

14a. *Columbia Glacier, Alaska #1*, 1962
PEN & INK & WATERCOLOR
5 1/2 x 8 1/8 inches

14b. *Columbia Glacier, Alaska #2*, 1962
PEN & INK & WATERCOLOR
5 1/2 x 8 1/8 inches

15. *Valdez Narrows, Alaska*, 1962
PEN & INK & WATERCOLOR
5 1/2 x 8 1/8 inches

16. *View of Arapahoe Glacier from Road to "Sunshine" and Gold Hill, Colorado*, 1964
PEN & INK & WATERCOLOR
5 1/2 x 8 1/4 inches

17a. *Sand Dunes National Monument, Colorado*, 1964
PEN & INK & WATERCOLOR
5 1/2 x 8 1/2 inches

17b. *Sand Dunes, Colorado*, 1964
PEN & INK & WATERCOLOR
5 1/2 x 8 1/2 inches

18. *First Snows of September, La Sal Mts., Moab, Utah*, 1964
PEN & INK & WATERCOLOR
5 1/2 x 8 1/4 inches

19. *Blue Valley Overlook, Near Highlands, NC*, n.d. (ca. 1964)
WATERCOLOR
4 15/16 x 7 1/16 inches

20. *Little Indian Garden at Sippewisset, MA*, 1966
PEN & INK & WATERCOLOR
5 1/2 x 8 7/16 inches

21. *Bermuda*, 1966
PEN & INK & WATERCOLOR
5 1/2 x 8 3/8 inches

22. *View North on Beach, Key Biscayne, FL*, 1967
PEN & INK & WATERCOLOR
8 3/16 x 10 15/16 inches

23. *Entrance to Cave Creek Canyon, Portal, Arizona*, 1968
WATERCOLOR
8 3/16 x 10 15/16 inches

24. *Gate to Chinatown, San Francisco*, 1969
PEN & INK & WATERCOLOR
8 1/4 x 5 7/16 inches

25. *Golden Gate Bridge, San Francisco*, 1969
PEN & INK & WATERCOLOR
8 3/16 x 5 1/2 inches

26. *Musical Chairs—Gulls at Bodega Bay, California*, 1969
PEN & INK & WATERCOLOR
8 1/4 x 5 1/2 inches

27a. *Sapelo Dunes*, 1969
PEN & INK & WATERCOLOR
8 3/16 x 11 inches

27b. *Sapelo Beach*, 1986
LITHOGRAPH
10 7/8 x 15 5/16 inches

28. *Jupiter Lighthouse, Florida*, 1970
PEN & INK & WATERCOLOR
6 x 9 inches

29a. *Seal Cove Rocks, Oregon*, 1970
PEN & INK & WATERCOLOR
6 x 8 3/4 inches

29b. *San Pedro Point, Montara Beach, California*, 1970
PEN & INK & WATERCOLOR
6 x 8 3/4 inches

30. *Pemaquid Point, Maine*, n.d.
HEAVY PEN & INK & WATERCOLOR
8 x 9 3/4 inches

31. *View from Mel's Island, Lake of the Woods, Canada*, 1971
PEN & INK & WATERCOLOR
8 3/4 x 6 inches

32. *Brown Pelicans, Tague Bay, St. Croix*, 1972
PEN & INK & WATERCOLOR
6 x 8 7/8 inches

33. *Dead Horse Point, Near Moab, Utah*, 1972
PEN & INK & WATERCOLOR
6 x 8 11/16 inches

34a. *Monument Valley, Arizona, Navajo Tribal Headquarters*, 1972
PEN & INK & WATERCOLOR
6 x 9 inches

34b. *Valley of the Gods, Mexican Hat, Utah*, 1972
PEN & INK & WATERCOLOR
6 x 9 inches

35. *Monument Valley, Utah, Sunset as Seen from Goulding's Trading Post*, 1972
PEN & INK & WATERCOLOR
8 7/8 x 11 3/4 inches

36. *Flower Market, Delft, Holland*, 1974
PEN & INK & WATERCOLOR
9 x 6 inches

37. *Cap d'Erquy, Coast of Brittany, France*, 1975
PEN & INK & WATERCOLOR
6 x 8 3/4 inches

38a. *Clouds from Plane Window*, 1974
WATERCOLOR
6 x 9 inches

38b. *English Countryside as Seen from Plane*, 1974
PEN & INK & WATERCOLOR
6 x 8 7/8 inches

39a. *Snake River Canyon, Our Raft in the Foreground*, 1975
PEN & INK & WATERCOLOR
6 x 8 3/4 inches

39b. *Snake River Float Trip with the Nature Conservancy Board Breaking Camp in the Morning*, 1975
PEN & INK & WATERCOLOR
6 x 8 11/16 inches

40. *Fireweed and Fish Traps, New Brunswick, Canada*, 1975
PEN & INK & WATERCOLOR
9 x 12 inches

41. *Near Oven Head, New Brunswick*, 1975
PEN & INK & WATERCOLOR
5 1/8 x 7 1/8 inches

42a. *Sapelo, Marsh-side, Low Tide*, n.d. (ca. 1976)
PEN & INK & WATERCOLOR
5 1/2 x 8 1/4 inches

42b. *Sapelo Marsh (Fall)*, n.d.
LITHOGRAPH

43a. *Valona, Georgia (Sapelo Island in Distance)*, 1978
PEN & INK & WATERCOLOR
7 x 11 1/16 inches

43b. *Shrimp Boats, Fort Myers, FL*, n.d.
PEN & INK & WATERCOLOR
8 3/4 x 12 inches

44. *Viceroy Butterflies, Valona, GA*, 1978
PEN & INK & WATERCOLOR
6 x 8 11/16 inches

45. *Yellow Bark Acacia Trees Near Nairobi, Kenya*, 1979
PEN & INK & WATERCOLOR
6 x 8 3/4 inches

46. *Crocodile Bar, Samburu, Kenya*, 1979
PEN & INK & WATERCOLOR
6 x 8 11/16 inches

47a. *African Buffalo and Warthogs as Seen from Treetops*, 1979
PEN & INK & WATERCOLOR
6 x 8 5/8 inches

47b. *Olive Baboon, Mother Protecting Baby in Misty Rain, Treetops, Kenya*, 1979
PEN & INK & WATERCOLOR
8 3/4 x 6 inches

48. *Cache la Poudre River, Colorado*, 1979
WATERCOLOR
8 3/4 x 6 inches

49. *Mt. Crested Butte, Ski Area*, 1979
PEN & INK & WATERCOLOR
8 3/4 x 6 inches

50. *Edge of Gunnison, Colorado*, 1979
PEN & INK & WATERCOLOR
6 x 8 7/8 inches

51. *Blue Ridge Parkway*, 1979
WATERCOLOR
8 3/4 x 6 inches

52. *Wetland Near Chattahoochee Nature Center*, 1979
PEN & INK & WATERCOLOR
8 11/16 x 6 inches

53. *St. Simons Island, GA*, 1980
WATERCOLOR
6 x 9 inches

54. *American River, Near Folsom, California*, 1980
PEN & INK & WATERCOLOR
6 x 8 3/4 inches

55. *View from the Drake Hotel, Chicago, Lakefront*, 1980
PEN & INK & WATERCOLOR
6 x 9 inches

56. *Near Whale Cove, Cape Breton, Nova Scotia*, 1981
PEN & INK & WATERCOLOR
6 x 8 15/16 inches

57. *Athens City Hall Eagle*, 1983
PEN & INK & WATERCOLORr
8 3/4 x 6 inches

58. *Old Campus, University of Georgia, Late Spring*, n.d.
PEN & INK & WATERCOLOR
7 5/8 x 11 5/8 inches

59. *Mountain Waterfall, Georgia*, n.d.
WATERCOLOR

60. *Purple Rhododendron*, n.d.
WATERCOLOR
10 x 14 1/8 inches

61. *Wild Iris-Cabo de São Vicente Near Sagres, Portugal*, 1985
WATERCOLOR
9 1/4 x 12 1/4 inches

62. *Aravaipa Canyon, Arizona*, 1993
WATERCOLOR
4 7/8 x 6 3/4 inches

about the AUTHOR

Dr. Eugene P. Odum is known as the father of modern ecosystem ecology. Rather than focusing in on a single facet of a given ecosystem, he advocates looking at the big picture, including the importance of the human element. Author of the first major textbook on ecology (in 1953), recipient of Sweden's Craaford Prize and the French government's Prize of the Institut de la Vie, and founder of the University of Georgia's Institute of Ecology, many consider him the foremost ecologist today. He and his wife, Martha, spent nearly sixty years globe-trotting; while he attended ecological conferences, she would illustrate the surrounding region in watercolor. Martha influenced the cultural life of Athens for more than fifty years, both through her own talents and through her support of others' talents. Together, they formed the ideal partnership between art and ecology.